T0070307

Billionaire
Success 2

JAVONTE' JENNINGS

authorHOUSE®

AuthorHouse™
1663 Liberty Drive
Bloomington, IN 47403
www.authorhouse.com
Phone: 1 (800) 839-8640

© 2017 Javonte' Jennings. All rights reserved.

No part of this book may be reproduced, stored in a retrieval system, or transmitted
by any means without the written permission of the author.

Published by AuthorHouse 02/22/2017

ISBN: 978-1-5246-7012-2 (sc)
ISBN: 978-1-5246-7011-5 (e)

Print information available on the last page.

Any people depicted in stock imagery provided by Thinkstock are models,
and such images are being used for illustrative purposes only.
Certain stock imagery © Thinkstock.

This book is printed on acid-free paper.

Because of the dynamic nature of the Internet, any web addresses or links contained in this book may have changed
since publication and may no longer be valid. The views expressed in this work are solely those of the author and do
not necessarily reflect the views of the publisher, and the publisher hereby disclaims any responsibility for them.

Contents

THANKS TO

* GOD

* FRINEDS

* FAMILY

DAILY COMMITMENT

WWW._____.COM

NAME OF OWNER:_____

PROPERTY ADDRESS:_____

PARCEL ID#:_____

BLOCK:_____

TAXES:_____

VALUE:_____

ADVERTISED PRICE:_____

OFFER AMOUNT:_____

NEAREST HOME IMPROVEMENT STORE

_____

_____

_____

_____

DAILY COMMITMENT

WWW._____.COM

NAME OF OWNER:_____

PROPERTY ADDRESS:_____

PARCEL ID#:_____

BLOCK:_____

TAXES:_____

VALUE:_____

ADVERTISED PRICE:_____

OFFER AMOUNT:_____

NEAREST HOME IMPROVEMENT STORE

_____

_____

_____

_____

DAILY COMMITMENT

WWW._____.COM

NAME OF OWNER:_____

PROPERTY ADDRESS:_____

PARCEL ID#:_____

BLOCK:_____

TAXES:_____

VALUE:_____

ADVERTISED PRICE:_____

OFFER AMOUNT:_____

NEAREST HOME IMPROVEMENT STORE

_____

_____

_____

_____

DAILY COMMITMENT

WWW._____.COM

NAME OF OWNER:_____

PROPERTY ADDRESS:_____

PARCEL ID#:_____

BLOCK:_____

TAXES:_____

VALUE:_____

ADVERTISED PRICE:_____

OFFER AMOUNT:_____

NEAREST HOME IMPROVEMENT STORE

_____

_____

_____

_____

DAILY COMMITMENT

WWW._____.COM

NAME OF OWNER:_____

PROPERTY ADDRESS:_____

PARCEL ID#:_____

BLOCK:_____

TAXES:_____

VALUE:_____

ADVERTISED PRICE:_____

OFFER AMOUNT:_____

NEAREST HOME IMPROVEMENT STORE

_____

_____

_____

_____

DAILY COMMITMENT

WWW._____.COM

NAME OF OWNER:_____

PROPERTY ADDRESS:_____

PARCEL ID#:_____

BLOCK:_____

TAXES:_____

VALUE:_____

ADVERTISED PRICE:_____

OFFER AMOUNT:_____

NEAREST HOME IMPROVEMENT STORE

_____

_____

_____

_____

DAILY COMMITMENT

WWW._____.COM

NAME OF OWNER:_____

PROPERTY ADDRESS:_____

PARCEL ID#:_____

BLOCK:_____

TAXES:_____

VALUE:_____

ADVERTISED PRICE:_____

OFFER AMOUNT:_____

NEAREST HOME IMPROVEMENT STORE

_____

_____

_____

_____

DAILY COMMITMENT

WWW._____.COM

NAME OF OWNER:_____

PROPERTY ADDRESS:_____

PARCEL ID#:_____

BLOCK:_____

TAXES:_____

VALUE:_____

ADVERTISED PRICE:_____

OFFER AMOUNT:_____

NEAREST HOME IMPROVEMENT STORE

_____

_____

_____

_____

DAILY COMMITMENT

WWW._____.COM

NAME OF OWNER:_____

PROPERTY ADDRESS:_____

PARCEL ID#:_____

BLOCK:_____

TAXES:_____

VALUE:_____

ADVERTISED PRICE:_____

OFFER AMOUNT:_____

NEAREST HOME IMPROVEMENT STORE

_____

_____

_____

_____

DAILY COMMITMENT

WWW._____.COM

NAME OF OWNER:_____

PROPERTY ADDRESS:_____

PARCEL ID#:_____

BLOCK:_____

TAXES:_____

VALUE:_____

ADVERTISED PRICE:_____

OFFER AMOUNT:_____

NEAREST HOME IMPROVEMENT STORE

_____

_____

_____

_____

DAILY COMMITMENT

WWW._____.COM

NAME OF OWNER:_____

PROPERTY ADDRESS:_____

PARCEL ID#:_____

BLOCK:_____

TAXES:_____

VALUE:_____

ADVERTISED PRICE:_____

OFFER AMOUNT:_____

NEAREST HOME IMPROVEMENT STORE

_____

_____

_____

_____

DAILY COMMITMENT

WWW._____.COM

NAME OF OWNER:_____

PROPERTY ADDRESS:_____

PARCEL ID#:_____

BLOCK:_____

TAXES:_____

VALUE:_____

ADVERTISED PRICE:_____

OFFER AMOUNT:_____

NEAREST HOME IMPROVEMENT STORE

_____

_____

_____

_____

DAILY COMMITMENT

WWW._____.COM

NAME OF OWNER:_____

PROPERTY ADDRESS:_____

PARCEL ID#:_____

BLOCK:_____

TAXES:_____

VALUE:_____

ADVERTISED PRICE:_____

OFFER AMOUNT:_____

NEAREST HOME IMPROVEMENT STORE

_____

_____

_____

_____

DAILY COMMITMENT

WWW._____.COM

NAME OF OWNER:_____

PROPERTY ADDRESS:_____

PARCEL ID#:_____

BLOCK:_____

TAXES:_____

VALUE:_____

ADVERTISED PRICE:_____

OFFER AMOUNT:_____

NEAREST HOME IMPROVEMENT STORE

_____

_____

_____

_____

DAILY COMMITMENT

WWW._____.COM

NAME OF OWNER:_____

PROPERTY ADDRESS:_____

PARCEL ID#:_____

BLOCK:_____

TAXES:_____

VALUE:_____

ADVERTISED PRICE:_____

OFFER AMOUNT:_____

NEAREST HOME IMPROVEMENT STORE

_____

_____

_____

_____

DAILY COMMITMENT

WWW._____.COM

NAME OF OWNER:_____

PROPERTY ADDRESS:_____

PARCEL ID#:_____

BLOCK:_____

TAXES:_____

VALUE:_____

ADVERTISED PRICE:_____

OFFER AMOUNT:_____

NEAREST HOME IMPROVEMENT STORE

_____

_____

_____

_____

DAILY COMMITMENT

WWW._____.COM

NAME OF OWNER:_____

PROPERTY ADDRESS:_____

PARCEL ID#:_____

BLOCK:_____

TAXES:_____

VALUE:_____

ADVERTISED PRICE:_____

OFFER AMOUNT:_____

NEAREST HOME IMPROVEMENT STORE

_____

_____

_____

_____

DAILY COMMITMENT

WWW._____.COM

NAME OF OWNER:_____

PROPERTY ADDRESS:_____

PARCEL ID#:_____

BLOCK:_____

TAXES:_____

VALUE:_____

ADVERTISED PRICE:_____

OFFER AMOUNT:_____

NEAREST HOME IMPROVEMENT STORE

_____

_____

_____

_____

DAILY COMMITMENT

WWW._____.COM

NAME OF OWNER:_____

PROPERTY ADDRESS:_____

PARCEL ID#:_____

BLOCK:_____

TAXES:_____

VALUE:_____

ADVERTISED PRICE:_____

OFFER AMOUNT:_____

NEAREST HOME IMPROVEMENT STORE

_____

_____

_____

_____

DAILY COMMITMENT

WWW._____.COM

NAME OF OWNER:_____

PROPERTY ADDRESS:_____

PARCEL ID#:_____

BLOCK:_____

TAXES:_____

VALUE:_____

ADVERTISED PRICE:_____

OFFER AMOUNT:_____

NEAREST HOME IMPROVEMENT STORE

_____

_____

_____

_____

DAILY COMMITMENT

WWW._____.COM

NAME OF OWNER:_____

PROPERTY ADDRESS:_____

PARCEL ID#:_____

BLOCK:_____

TAXES:_____

VALUE:_____

ADVERTISED PRICE:_____

OFFER AMOUNT:_____

NEAREST HOME IMPROVEMENT STORE

_____

_____

_____

_____

DAILY COMMITMENT

WWW._____.COM

NAME OF OWNER:_____

PROPERTY ADDRESS:_____

PARCEL ID#:_____

BLOCK:_____

TAXES:_____

VALUE:_____

ADVERTISED PRICE:_____

OFFER AMOUNT:_____

NEAREST HOME IMPROVEMENT STORE

_____

_____

_____

_____

DAILY COMMITMENT

WWW._____.COM

NAME OF OWNER:_____

PROPERTY ADDRESS:_____

PARCEL ID#:_____

BLOCK:_____

TAXES:_____

VALUE:_____

ADVERTISED PRICE:_____

OFFER AMOUNT:_____

NEAREST HOME IMPROVEMENT STORE

_____

_____

_____

_____

DAILY COMMITMENT

WWW._____.COM

NAME OF OWNER:_____

PROPERTY ADDRESS:_____

PARCEL ID#:_____

BLOCK:_____

TAXES:_____

VALUE:_____

ADVERTISED PRICE:_____

OFFER AMOUNT:_____

NEAREST HOME IMPROVEMENT STORE

_____

_____

_____

_____

DAILY COMMITMENT

WWW._____.COM

NAME OF OWNER:_____

PROPERTY ADDRESS:_____

PARCEL ID#:_____

BLOCK:_____

TAXES:_____

VALUE:_____

ADVERTISED PRICE:_____

OFFER AMOUNT:_____

NEAREST HOME IMPROVEMENT STORE

_____

_____

_____

_____

DAILY COMMITMENT

WWW._____.COM

NAME OF OWNER:_____

PROPERTY ADDRESS:_____

PARCEL ID#:_____

BLOCK:_____

TAXES:_____

VALUE:_____

ADVERTISED PRICE:_____

OFFER AMOUNT:_____

NEAREST HOME IMPROVEMENT STORE

_____

_____

_____

_____

DAILY COMMITMENT

WWW._____.COM

NAME OF OWNER:_____

PROPERTY ADDRESS:_____

PARCEL ID#:_____

BLOCK:_____

TAXES:_____

VALUE:_____

ADVERTISED PRICE:_____

OFFER AMOUNT:_____

NEAREST HOME IMPROVEMENT STORE

_____

_____

_____

_____

DAILY COMMITMENT

WWW._____.COM

NAME OF OWNER:_____

PROPERTY ADDRESS:_____

PARCEL ID#:_____

BLOCK:_____

TAXES:_____

VALUE:_____

ADVERTISED PRICE:_____

OFFER AMOUNT:_____

NEAREST HOME IMPROVEMENT STORE

_____

_____

_____

_____

DAILY COMMITMENT

WWW._____.COM

NAME OF OWNER:_____

PROPERTY ADDRESS:_____

PARCEL ID#:_____

BLOCK:_____

TAXES:_____

VALUE:_____

ADVERTISED PRICE:_____

OFFER AMOUNT:_____

NEAREST HOME IMPROVEMENT STORE

_____

_____

_____

_____

DAILY COMMITMENT

WWW._____.COM

NAME OF OWNER:_____

PROPERTY ADDRESS:_____

PARCEL ID#:_____

BLOCK:_____

TAXES:_____

VALUE:_____

ADVERTISED PRICE:_____

OFFER AMOUNT:_____

NEAREST HOME IMPROVEMENT STORE

_____

_____

_____

_____

STORAGE UNITS IDEAS

BILLS PAID IN ADVANCE

<u>PICK YOUR DAYS WISLEY</u>

SCHUDULE YOUR TIME WITH A CALENDAR

BE NEAT

SHELVES

STORE ONLY YOUR GOODS/ NO THEFT

LOG IN & LOG OUT

THIS IS THE "PREP STATION" FOR YOUR RANCH

TRANSPORTATION

PLANE

WWW._____.COM

FLIGHT COST:_____

DATE:_____

TERMINAL:_____

30 MINUTE WAIT……….. SECURITY CHECK TIME

DEPART TIME:_____

CONNECTION IN:_____

DELAY TIME:_____

TERMINAL:_____

NEAREST STORAGE FACILITIES

_____

_____

_____

_____

PLANE

WWW._____.COM

FLIGHT COST:_____

DATE:_____

TERMINAL:_____

30 MINUTE WAIT……….. SECURITY CHECK TIME

DEPART TIME:_____

CONNECTION IN:_____

DELAY TIME:_____

TERMINAL:_____

NEAREST STORAGE FACILITIES

_____

_____

_____

_____

PLANE

WWW._____.COM

FLIGHT COST:_____

DATE:_____

TERMINAL:_____

30 MINUTE WAIT……….. SECURITY CHECK TIME

DEPART TIME:_____

CONNECTION IN:_____

DELAY TIME:_____

TERMINAL:_____

NEAREST STORAGE FACILITIES

_____

_____

_____

_____

PLANE

WWW._____.COM

FLIGHT COST:_____

DATE:_____

TERMINAL:_____

30 MINUTE WAIT……….. SECURITY CHECK TIME

DEPART TIME:_____

CONNECTION IN:_____

DELAY TIME:_____

TERMINAL:_____

NEAREST STORAGE FACILITIES

_____

_____

_____

_____

PLANE

WWW._____.COM

FLIGHT COST:_____

DATE:_____

TERMINAL:_____

30 MINUTE WAIT……….. SECURITY CHECK TIME

DEPART TIME:_____

CONNECTION IN:_____

DELAY TIME:_____

TERMINAL:_____

NEAREST STORAGE FACILITIES

_____

_____

_____

_____

PLANE

WWW._____.COM

FLIGHT COST:_____

DATE:_____

TERMINAL:_____

30 MINUTE WAIT……….. SECURITY CHECK TIME

DEPART TIME:_____

CONNECTION IN:_____

DELAY TIME:_____

TERMINAL:_____

NEAREST STORAGE FACILITIES

_____

_____

_____

_____

PLANE

WWW._____.COM

FLIGHT COST:_____

DATE:_____

TERMINAL:_____

30 MINUTE WAIT……….. SECURITY CHECK TIME

DEPART TIME:_____

CONNECTION IN:_____

DELAY TIME:_____

TERMINAL:_____

NEAREST STORAGE FACILITIES

_____

_____

_____

_____

PLANE

WWW._____.COM

FLIGHT COST:_____

DATE:_____

TERMINAL:_____

30 MINUTE WAIT……….. SECURITY CHECK TIME

DEPART TIME:_____

CONNECTION IN:_____

DELAY TIME:_____

TERMINAL:_____

NEAREST STORAGE FACILITIES

_____

_____

_____

_____

PLANE

WWW._____.COM

FLIGHT COST:_____

DATE:_____

TERMINAL:_____

30 MINUTE WAIT……….. SECURITY CHECK TIME

DEPART TIME:_____

CONNECTION IN:_____

DELAY TIME:_____

TERMINAL:_____

NEAREST STORAGE FACILITIES

_____

_____

_____

_____

CAR

NEAREST CAR AUCTION LOCATIONS

_____(_____)

_____(_____)

_____(_____)

_____(_____)

DATE OF AUCTION:_____

TIME AUCTION STARTS:_____

AMOUNT OF CARS ADVERTISED:_____

BUS

BUS STOP LOCATION:_____

TIME SCHDULED:_____

TRANSIT PHONE#:_____

BUS #:_____

BUS

BUS STOP LOCATION:_____

TIME SCHDULED:_____

TRANSIT PHONE#:_____

BUS #:_____

BUS

BUS STOP LOCATION:_____

TIME SCHDULED:_____

TRANSIT PHONE#:_____

BUS #:_____

BUS

BUS STOP LOCATION:_____

TIME SCHDULED:_____

TRANSIT PHONE#:_____

BUS #:_____

BUS

BUS STOP LOCATION:_____

TIME SCHDULED:_____

TRANSIT PHONE#:_____

BUS #:_____

BUS

BUS STOP LOCATION:_____

TIME SCHDULED:_____

TRANSIT PHONE#:_____

BUS #:_____

Printed in the United States
By Bookmasters